Yamile Venticuaba Hernández
SILVA Miranda Tchikete

Technological Innovation In Agricultural Processes

Yamile Venticuaba Hernández
SILVA Miranda Tchikete

Technological Innovation In Agricultural Processes

Technological Innovation in Agricultural Processes in the Town of Oshomucuyo- Ondjiva

ScienciaScripts

Imprint

Cover image: www.ingimage.com

This book is a translation from the original published under ISBN 978-3-639-68682-1.

Publisher:
Sciencia Scripts
is a trademark of
Dodo Books Indian Ocean Ltd. and OmniScriptum S.R.L publishing group

120 High Road, East Finchley, London, N2 9ED, United Kingdom
Str. Armeneasca 28/1, office 1, Chisinau MD-2012, Republic of Moldova, Europe
Managing Directors: Ieva Konstantinova, Victoria Ursu
info@omniscriptum.com

Printed at: see last page
ISBN: 978-620-8-63589-3

TECHNOLOGICAL INNOVATION IN AGRICULTURAL PROCESSES IN THE TOWN OF OSHOMUCUYO-ONDJIVA

AUTHORS

MSc. YAMILE VENTICUABA HERNÁNDEZ ENG. SILVA MIRANDA TCHIKETE

FEBRUARY 2024

"Poets are born poets, they are not made. I'm not a poet, but am a botanist, so I offer the fruit of the small harvest that God has given me"

= Carl Von Linne=

SUMMARY

There is broad consensus on the importance of science and technology for the development of contemporary societies. However, little progress has been made in concretely evaluating the impact of science and technology on local development. This research aims to provide an assessment by showing the general effects of knowledge of technological innovation on the growth of families' quality of life. The aim of the research was to develop actions that contribute to innovative farm management in the locality of Oshomucuyo, Ondjiva. The work took place between March and September 2023. The design was structured in phases that gave chronological and systemic output to the specific objectives. The research was qualitative and based on the principles of participatory action research, as described by Roa et al. (2002). The historical-logical, analytical-synthetic and systemic-structural methods were used. Research was the fundamental strategy of the investigation, which was qualitative in terms of gathering data and quantitative in terms of interpreting it. The research results were analysed using tables and graphs in order to make them easier to understand. The information was analysed using the Excel package and Word for Windows version 10. A proposal was drawn up for technological innovation actions in the conditions of the region, which influence the improvement of agricultural development in the locality of Oshomucuyo. Ondjiva.

***Keywords**: innovation, technology, improvement, actions.*

CONTENTS

I. INTRODUCTION

Agricultural production can contribute to sustainable development in different ways: socially, by developing the production of nutritious and harmless food at reasonable prices, which generates employment, reduces health risks and poverty. Environmentally, by making efficient use of all renewable and non-renewable resources, reducing the loss of agrochemicals through percolation, volatilisation and erosion, maintaining or improving soil quality, and minimising the risk of water contamination and greenhouse gas emissions into the atmosphere. Economically, by generating wealth and promoting trade in foodstuffs (Lal, 2004).

According to Franco (2000), knowledge management has recently been seen as one of the social factors that not only contributes to the development of production, but also to local development. The focus on local development also includes agricultural development, urban development, industrial development, community development, etc., but does not remove the relative independence of each of these developments.

Arocena and Sutz (2002) say: "In certain societies, a considerable proportion of the population and a large number of organisations are involved, on a more or less permanent basis, in activities that demand both formal and tacit knowledge, require advanced education and offer opportunities to systematically increase the skills acquired, through various interaction processes, especially those aimed at innovation understood as problem-solving.

There is broad consensus on the central importance of science and technology for the development of contemporary societies. The available literature reflects this appreciation by showing the general effects of knowledge on economic growth. However, little progress has been made in concretely assessing the impact of science and technology on local development.

In sub-Saharan Africa (SSA), around 80 % of rural families earn their income by farming on small plots of land, averaging less than 2 ha, with production geared towards self-consumption and surplus. In this way, farmers' activities are integrated into the economy, contributing a third of Africa's Gross Domestic Product (GDP2). Despite the potential that agriculture has, we still realise that farmers face enormous challenges, including food insecurity, adapting to changing economic conditions and climate change, technology not suited to the local reality and overcoming poverty (Marassiro, Romarco de Oliveira, and

Pereira, 2021).

Among the various reasons that can be deduced, the insufficient understanding of the intermediary mechanisms between the production of knowledge and the development of society has undoubtedly delayed the production of impact indicators. There is also the criterion that science and technology must play an important role in tackling these provocations and transformations. But in order to do so, it is necessary to focus on them not exclusively as technical and intellectual processes, but as social processes, in their social conditioning and impact (Núñez, 2019).

The current state of knowledge on these or other problems linked to social development does not always make it possible to identify the intervention of causally relevant factors. Often, and as the problem becomes more complex (poverty, for example), only co-variation relationships can be firmly established. This situation would not inhibit the possibility of continuing to think about causal relationships, especially along a line of enquiry that is more in-depth and of a length that we like. But it does suggest that it would be advisable to provisionally work on the hypothesis of correlation, until we have a better understanding of the impact of science and technology on society.

The environment demands that organisations, as a strategy for their survival and development, foster innovative practices that have an impact on the creation of products and services that are the result of knowledge and the management of people in permanent harmony with the environment. Faced with this reality, there is a need to adopt conduct that encourages the development of the creative and innovative capacities of the organisation and its members (Te serre, 2010).

The analysis of MATRE's annual balance sheet information (2020) reflects the prevalence of problems that have been identified for several years, and which even have not been solved or have only covered part of the problem. These stand out for their transcendence:

1. Insufficient comprehensive preparation at all levels for the development of strategic planning of agricultural processes with a sustainable approach.
2. Lack of resources for the efficient development of production processes.
3. Poor organisation, control, monitoring and evacuation of production activity, as well as the process of innovation and technology transfer.
4. Poor exploitation of natural resources (soil and water being the most important).
5. Use of obsolete technologies with very little agroecological support.

6. Little motivation to carry out research to improve processes. Among others.

For all these reasons, it is necessary to start establishing new indicators of the impact of science on local development that more accurately reflect the evolution of the different territories in their socio-economic development:

Scientific problem: shortcomings in innovation management in the town of Oshomucuyu have an impact on agricultural development in the area.

We hypothesise that the use of technological innovation on farms could increase production levels in the town of Oshomucuyu, Ondjiva.

General objective: to develop actions that contribute to innovative farm management in the locality of Oshomucuyu, Ondjiva

Specific objectives:

1. Theoretically ground the general elements of technological innovation on the farm.
2. To diagnose the current situation of technological innovation systems in agricultural and livestock processes in the locality of Oshomucuyu. Ondjiva.
3. Identify the main problems facing the agricultural production systems under study, using the SWOT Matrix.
4. Designing technological innovation actions in the conditions of the region, which influence improvement of agricultural development in the locality of Oshomucuyu. Ondjiva.

II. REVIEW BIBLIOGRAPHY

2.1. General about technology and innovation

According to Castro (2002), technology plays a critical role in a company's competitiveness and is one of the intangible factors that is most difficult to manage. The new scenario is identified with the acceleration of technological change and the shortening of product life cycles, hence the strategic importance of carrying out effective technology management in the company.

The same source defines technology as the means of transforming ideas into processes or services, making it possible to improve or develop processes. However, although its etymological roots reduce it to the science of industrial arts, it does not consist solely of methods, machines, procedures, instruments, programming methods, materials and teams that can be bought and exchanged, but is also a state of mind, the expression of a creative talent and the ability to systematise knowledge so that it can be used by society as a whole.

According to Varrer (2010), innovation is defined as the transformation of an idea into a saleable, new or improved product or team; into an operating process in industry or commerce, or into a new methodology for social organisation. This author's judgement on all the scientific, technical, commercial and financial steps necessary for the successful development and commercialisation of the new or improved product, process or social service. The act by which a technological change is introduced for the first time in an organisation or company is called innovation.

Castro (2002) defines technological innovation as the set of scientific, technological, financial and commercial activities that make it possible:

- Introducing new or improved products on the domestic or foreign market (examples: medicines, plant protection products, equipment, technological devices for different purposes)
- Introducing new or improved services (examples: new preventive, care and/or assistance services).
- Implementing new or improved production processes or procedures (examples: medical-surgical, teaching-educational, information and automation).
- Introducing and validating new or improved management techniques and organisational systems that are used in our factories and companies.

Therefore, according to Pineda (2010), technological innovation comprises new products and processes and significant changes, from a technological point of view, in products and processes. An innovation is understood to have been applied when it is put on the market (product innovation) or used in a production process (process innovation).

2.2. Management of innovation

According to the definitions of technological innovation given in the previous paragraphs, we can identify the set of elements that are considered necessary for a company to efficiently manage the innovation process: human capital, knowledge, entrepreneurial spirit, co-operation and an innovative culture.
Pineda (2006) considers the following stages in the innovation process:

1. Starting from a problematic situation (opportunity).
2. A diagnosis is made based on the maintenance carried out.
3. Possible solutions (ideas) are presented
4. We proceed to realise:

a) The repair

b) Adaptation or

c) Innovation

5. New technology is generated
6. The new technology is implemented
7. The evaluation is carried out
8. Note the innovation

It can be concluded that technological innovation starts from a need to solve a problem and requires the knowledge and experience of the individual to solve it, which sets in motion the process of generating innovation, resulting in an improvement in production systems in many cases.Suarez et al. (2007) recognise the following as the main factors hindering innovation:
ECONOMIC FACTORS

- Excessive risks
- Lack of financial resources
- High innovation costs
- Prolonged and uncertain recovery period

TECHNICAL AND PRODUCTION FACTORS

- Insufficient infrastructure for innovation
- Difficulties in obtaining raw materials and materials
- Difficulty obtaining replacements
- Difficulty in technical and production maintenance
- Insufficient marketing application
- Insufficient level of technological co-operation with:

✓ Other companies

✓ Universities and research centres

✓ Consultancy

✓ Other technology management organisations

✓ Insufficient level of technological cooperation FACTORS RELATING TO HUMAN RESOURCES

- Lack of qualified staff
- Resistance to change
- Insufficient capacity building and technological training
- Insufficient capacity of directors in technological management FACTORS RELATING TO INFORMATION
- Insufficient information on technological advances
- Insufficient market information.
- Insufficient information about competitors.
- Insufficient dissemination in company areas OTHER FACTORS
- Limited technological opportunities
- Limitations of current legislation and regulations
- Difficult to commercialise new products
- Lack of technological dynamism in the industry.

On the other hand, other authors claim that there are three factors that limit successful innovation:

- Personnel: you will have to take into account which people are needed in technical development, and what management can be carried out to maximise their joint productivity.

- Structure: an innovative organisation requires adequate technical and market information, and also needs its results to be integrated within the general objectives of the organisation and that they are adequately transferred to the end user.
- Strategy: this has to include aspects of strategic planning and implementation at two levels: general (for the whole organisation) and particular (focused on the unit responsible for developing and acquiring innovation).

2.3. Importance of the innovation process technology

Given the context in which organisations are currently competing, innovation is becoming a daily requirement. Authors such as Tushman and Ó Reilly (2002) consider that organisations that want to be successful today and tomorrow must be able to manage different innovations at , as they will be incremental because they are mature products that are in their retention phase, but also discontinuous, which may even be products or markets that displace today's successful ones. However, not all companies have the conditions required to generate a sufficient number of ideas, nor for them to successfully materialise into an improvement in products/services, processes and/or management systems.

Technological innovation is an important source of increased productivity and is currently considered synonymous with modernity (Castro, 2002). It also makes it possible to obtain a greater volume of goods and services, improve quality and introduce new marketing methods; through greater automation and information technology, it relieves workers of heavy and monotonous tasks. Automation can also improve material handling, storage, communication systems and quality control. However, technological factors alone do not lead to effective results if they are not accompanied by the development of the human factor.

According to the ideas of Bohorquez (2003); Perez (2003) and Socorro (2003), technology is not the only factor that determines competitiveness, although today the criterion is widely held that among all the things that can change the rules of competence, technological change is the most prominent. Competitive advantages today derive from scientific knowledge converted into technologies.

Since the mid-1980s, the technological factor has become a strategic vector that allows companies to improve their competitive position, since its absence leads to a serious lack of capacity to generate innovations in products and processes. It is necessary to manage these technological resources as efficiently as the others so that the company acquires a greater capacity to adapt and the ability to anticipate, and even cause disruptions that allow them to renew their competitive advantages at the right time (Nevado, 2003).

Innovation is the key element that explains competitiveness. Socorro (2003) is of the opinion that "The competitiveness of a nation depends on the capacity of its industry to innovate and improve. The company achieves competitive advantage through innovation." Chenais quoted by Socorro (2003) emphasised that: "Innovative activity actually constitutes, with human capital, one of the main factors determining the competitive advantage of advanced industrial economies".It is clear that innovation ends with the successful introduction of the product onto the market. The close connection between the current concept of competitiveness and innovation is clear: saying that new products must be successful is the same as saying that they must be competitive.

According to Escobar (2010), in today's environment it is necessary to develop human resources, information systems and technological capabilities in line with the new challenges. Hence the importance of the innovation process. This implies the renewal and expansion of processes, products and services, changes in management organisation and changes in the qualifications of human capital. However, it should not be understood as a purely technical concept, but has economic and social roots, and its analysis requires an understanding of the two essential characteristics:

- The aim of innovation is to exploit the opportunities offered by change, which means that it is fundamental in generating an innovative culture that allows the company to be able to adapt to the new situations and demands of the market in which it competes.
- The innovative character is based on the complexity of the technological research process and the unpredictable changes that move the market and the competence itself.

According to Carvalho et al. (2011), innovation can provide:

- Increased demand for their products and services with the creation of new markets, clear differentiation from competitors and increased perceived quality;
- Better defence of its competitive position through products and services that are highly difficult to imitate;
- Reducing costs by improving the efficiency of your production and management processes;
- Expanding margins with high value-added products and services that allow them to offer differentiated premium prices;
- Increased competence to innovate through the practice of launching

innovative products and services, which leads the company to increase its ability, volume of knowledge and attitude towards innovation. Over time, with further training and qualification, better and faster market launches can occur.

2.4. Production systems

The agricultural system refers to the regional organisation of the various plant and animal production systems, which takes into account the peculiarities and similarities of these different systems. According to Hirakuri et al. (2012), this organisation should allow for the construction of production models and arrangements that describe the predominant production systems in the region as accurately as possible. The production system is made up of all the cultivation or breeding systems within a rural property, defined on the basis of production factors (land, capital and labour) and interconnected by a management process.

An agricultural production system is made up of production subsystems such as the annual crop production subsystem and the perennial crop production subsystem. In this production system, the farm cannot ignore the human component or rational entity, the farmer and his family. It is the farmer who makes production, resource and energy use decisions (MINADER, 2020). The farm as a unit of operation can be seen as an agricultural production system in which resources, inputs, agents interact as transformers managed by the operator and under constraints of physical, biological, social and economic environments, to contribute to the overall agricultural product.

For this source, any concept of an agricultural production system must include at least the following:

a) A purpose for which the system was designed;

b) Borders: some procedure to say what is inside and what is outside the system;

c) Context: the external environment in which the system operates;

d) Components: main components that are related in the formation of the system;

e) Interactions between components;

f) Resources: internal system components;

g) Contributions used by the system, but coming from outside it;

h) Main products or desired achievements;

i) Useful, albeit incidental, by-products.

According to Astier, Masera, and Galván (2008), the subsystem approach differs substantially from the arbitrary consideration of parts of the system. Firstly, it must be recognised that any part of a system can be considered as another system in itself, examined in the same way and form.

2.5. Angolan agriculture: constraints, objectives and challenges

Lote (2015) points out that the agricultural sector in Angola falls under the Ministry of Agriculture (MINAGRI), the Angolan state body that manages agricultural policies in coordination and with the help of other partners, both international and national, who have been collaborating with it to achieve the goals set out in the National Development Plan. It says that, together with its national partners and international organisations, it prepares the agricultural season each year for its economic agents, or target public, who are grouped into Family Farms (EAF), predominantly geared towards subsistence agricultural production; they have production systems geared towards self-sufficiency, but which progressively generate surpluses that are sold to informal markets in rural areas and to the main urban centres.

The author states that EAF production systems are complex, based on manual soil management practices, predominantly use local seeds and, although they have low productivity, are responsible for a great diversity of products. They are generally small in size and are based on the use of shifting cultivation methods, i.e. in the rainy season they farm areas located in the foothills of mountains and in the dry season on the banks of rivers; and Entrepreneurial farms (EAE), characterised by being totally geared towards the market and maximising available resources, i.e. production is essentially intended for marketing and also for making investments in the sector. They are based on poorly diversified production systems, which are often mechanised, and have higher average yields than EAFs, which make up the vast majority of farms in Angola.

The promotion of agriculture is fundamental when we think about establishing a new model of economic growth that is intended to be sustainable and, above all, less dependent on oil. The agricultural sector is important not only in terms of promoting self-sufficiency and food security, but also in terms of supplying raw materials for the processing industry and creating jobs, one of the crucial problems in rural areas, with a view to diversifying the economy. For the sustained growth of the agricultural sector, it is agriculture needs to evolve beyond the mere subsistence level (CEIC/UCAN, 2021).

The same literature adds that, as such, the main problems (obstacles) should be identified and solutions sought. The majority of the cultivated area is under the responsibility of family farms, with an average area per holding of 2.32 ha, using low-efficiency production techniques, being not very productive, making it difficult to create production surpluses because small family producers have little incentive to produce beyond the family's subsistence levels.

For Victorino Mauro (2018), it is believed that the agricultural sector represents a very real investment alternative for Angola in relation to an economy that is very dependent on oil. Based on the soil and climate conditions, intensive and extensive agricultural holdings could be the ones with the greatest potential and, to that extent, offer the greatest potential in terms of economic development. So, while it is true that the sector currently has important constraints, which are related to the fact that the Angolan economy is highly inflated and marked by great exchange rate instability, which has a very significant influence on sectors that very dependent on the foreign market, such as the agricultural sector, it is also true that in the past Angola was one of the world's largest exporters of coffee, cotton, sisal, maize, cassava and bananas, and from there the desire to regain this position is something that seems to be within reach.

The results of the 2018/19 campaign, reflected in the Ministry of Agriculture's report, show that the number of tractors is low and that they are only used in the business sector. The animal-drawn plough is used in both the family and business sectors, but with an average per holding of no more than 0.2 and 0.3 tools in the family and business sectors respectively. Some of the constraints on the development of Angolan agriculture are related to the low level of human capital and knowledge, the availability of financial resources and the lack of incentives for producers and technicians. The low level of human capital and knowledge involves aspects related to:

• Low yields due to the low level of technology and difficulties in accessing quality agricultural inputs in the necessary quantities and at the right time, such as quality seeds suited to the region, tools and equipment, etc.

tools suitable for working the land, animal and motorised traction, fertilisers, means of preventing and combating plant and animal pests and diseases;

• Lack of an effective rural extension service that can contribute to technological progress;

• Research is not very advanced, with a lack of investment in programmes and laboratories;

• Cultural barriers and a lack of knowledge on the part of farmers and technicians in technical, management, market and other terms, necessary for the projection of modern agriculture;

• Poor business management;

• High costs associated with importing inputs;

• Inexperience in operating in a market economy system on the part of the various authors, including the governing body;

With regard to the availability of financial resources, the main constraints relate to;

• Poor condition of roads, especially secondary and tertiary roads, and bridges, which hinders not only access to markets for the disposal of products, constituting a major obstacle to the commercialisation of agricultural and livestock products, but also access to the market for inputs;
• Lack or deterioration of other infrastructures, such as facilities for the different services, product storage and seed conservation;
• The absence of incentives for producers, service providers and technicians has the following consequences;
• Rural exodus, whereby new generations are increasingly disconnected from the land andunfamiliar with the techniques of working the land, making it difficult for them to return to the countryside;
• Growing feminisation of farming, due to the exodus of young people to urban centres;
• There are no rural financial services or they are very weak when they do exist;

• Possible land conflicts that could arise between family farmers and businessmen due to the effects of the latter's presence in unused areas;
• The population's low standard of living, which limits expectations and demand for consumer goods;

• Unattractive environment for business investment.

According to CEIC/UCAN (2019), resolving some of these problems is to some extent included in the government's objectives for promoting the integrated and sustainable socio-economic development of the agricultural sector, as set out in the 2025 programme. Among the various specific objectives defined, we can highlight the formulation of an agricultural policy, the development of agricultural research capacities, the implementation of a credit line to support associations, cooperatives, small and medium-sized producers, the rehabilitation of infrastructure to support productive activity, the promotion of trade in rural areas, the integrated promotion of strategic sectors, economic diversification and the promotion of public-private partnerships.

The source highlights the following as priority areas for intervention in terms of agrarian policy for the development of agriculture: the promotion of rural trade, technical assistance and agricultural extension services, agricultural credit and scientific research

2.6. Angola, Agriculture and Traditions

Almost the entire population is essentially rural, according to Simões (2019), there are different ethnic groups and, as such, with different socio-economic structures, Angola cannot be attributed to a single rural world, but to several. These rural areas are naturally associated with different types of economy. There are the peoples of the south-east, with an economy geared towards hunting and gathering, with almost no agricultural production whatsoever; the agricultural economy of the peoples of the north, with leftovers for barter, but no livestock; the pastoral economy of the peoples of the coast-south and south; and finally, the agro-pastoral economy of the peoples of the centre.

In the author's view, ecological constraints (climate, soil, etc.) and the response of men to these constraints in order to adapt better to the environment, will result in the distribution of the population in Angola.

An attempt was made to summarise traditional Angolan agriculture. For INE (2018), at this point, a brief comparison was made between the two sectors that characterise Angolan agriculture: the business sector and the traditional sector. Subsequently, the territory was divided into large agricultural regions, according to the population density of the areas and the type of predominant activity practised, agriculture or agro-pastoralism

In an article published by Rocha, Mora and Romero (2016), it is stated that the biggest problem with poor agricultural production is that most rural communities lack food security because they are unaware of the technical aspects managing their fields, such as bad farming practices. soil conservation (derived from low soil fertility, poor drainage); deforestation, (irregular rainfall and soil erosion), cultural, traditional aspects of farming, lack of knowledge, technical assistance in establishing other crops, lack of resources for purchasing inputs, lack of production alternatives and diversification in agriculture, all these factors limit production, putting their food security at risk.

The available resources are not well utilised, as they are used indiscriminately and in an irrational and unsustainable way. The majority of rural communities are vulnerable due to poor agricultural production (MINADER, 2022).

According to CEIC/UCAN (2021), the agricultural production system is made up of two subsystems, the entrepreneurial production subsystem and the peasant production subsystem. The entrepreneurial production subsystem is a subsystem formed by a minority sector that owns large tracts of arable land, also has the ability to hire moderate management and have technical and credit assistance, according to their needs. In terms of marketing, it eliminates all intermediaries for inputs and products and most of its production is destined for export. According to the same source, among the characteristics of the peasant production subsystem, its members make up a large mass of the population: predominantly farmers with small tracts of land, most of whom have no agricultural vocation. As for their production, most of it is destined for self-consumption or family subsistence. Although their financial and material resources are scarce or non-existent, they abuse them and exert unequal demographic control over natural resources, especially land. They also have little access to credit, very little or no available capital and insufficient or non-existent economic incentives, a low level of education, traditionalism and the absence of rural community organisations.

Republic of Angola (2020) states that if the means of production and the labour force belong to the worker, then the products of their labour belong to them. In simple commodity production, production is basically for consumption and not for exchange. The limited development of the productive forces and the social division of labour mean that market or peasant production only develops in an incipient, secondary and subordinate way.

2.7. Development of family farming at Angola

According to the researcher Prevedel (2023), family farming is carried out by families, with production geared towards their subsistence. This activity takes place on small properties and is cultivated by family labour. Normally, family farming is more sustainable. In addition to these characteristics, family farming also has:

- Greater productive diversity;
- Low use of agricultural inputs;
- Low use of technology.

The same author adds that the definition of family farming for developed countries is different from that for developing countries, such as Angola, since in the former case this type of farming is integrated into the market and is dedicated to the production of marketable products. As such, the family farming referred to in this article is usually referred to as peasant or traditional farming. This term is used to identify the social class that lives in the countryside, from the countryside and for the countryside and that practises essentially subsistence farming, only selling surpluses on local markets. The inclusion of this social group is fundamental to characterising Angolan agriculture today.

Neto (2008) highlights five essential features that characterise peasant society: its complete autonomy from the surrounding society, which dominates it but tolerates its originalities; the importance of the structural organisation of the family group in organising its economic and social life; despite this society's relationship with the surrounding society they live in a society that is self-sufficient in economic terms; relations with neighbouring collectivities are weak, with internal relations of inter-knowledge being much stronger; and those responsible for measuring between peasant collectivities and the surrounding society play a fundamental role.

In Angola, according to Pedro (2008), agriculture is fundamentally of the family type, where the use of advanced technology is scarce and manual tools are normally used for land preparation, weeding, sowing and harvesting. The use of agricultural inputs is low and yields are usually low when compared to world production. According to MINADER (2004), it is estimated that traditional (or peasant) farmers account for 80 per cent of national production. Of the remaining production, 18 per cent comes from medium-sized farmers, and only 2 per cent from commercial farmers.

2.8. Main characteristics of agricultural systems in Cunene and managing bodies

In Cunene, according to Ribeiro et al. (2021), massango, massambala and rainfed maize are the main crops, and because these are essentially pastoralist communities, agricultural issues are to a certain extent relegated to the background, which is why, at the family level, the threshold of 1 hectare per crop is hardly exceeded. In Huíla, on the other hand, there are notable differences, starting with the way in which the land is occupied, with not only rainfed land par excellence (ongongo), but also lowland (onaka) and the intermediate zone (ombanda), a strategy that allows farming at different times of the year and contributes to greater stability for households in rural areas. In addition, there is excellent use of cattle in Huíla, both in the traction of agricultural implements and in the use of manure to improve soil fertility, unlike

Cunene. In the author's opinion, agriculture and pastoralism are the activities of choice in the region, although with different incidences. In terms of pastoralism, cattle, goats, sheep, pigs and poultry predominate throughout the region, as well as donkeys and horses, the latter two being used mainly as pack and transport animals. Livestock production is purely extensive, with free grazing for most of the year, with the exception of the months when crops are grown. In some areas, free grazing during the dry season has been a source of conflict due to the agricultural utilisation of river banks. The same source says that between the residential centre and the watering point, as the dry season progresses, usually after the second half of September, the situation becomes untenable, forcing the herds to move to areas where they can find pasture and water (transhumance). This displacement, which is an ancestral practice, is mostly carried out by groups of farmers from the same area, involving not only cattle but also goats and sheep.The table lists the main traditional practices and techniques in use in family farming and livestock production in the context of climate change mitigation.

Table 1. Practices and techniques adopted to mitigate climate change

PROVÍNCIA	PRÁTICA OU TÉCNICA ADOPTADA PARA A MITIGAÇÃO DE ALTERAÇÕES CLIMÁTICAS	
	AGRICULTURA	PECUÁRIA
CUNENE	• Cultura de massango e massambala, dada a sua resistência a seca; • Implantação de pequenas hortas em áreas com alguma humidade, especialmente em margens de cursos de água; • Adopção de variedades de milho de ciclo curto. • Uso de Omo misturado com gasóleo na luta contra a broca do colmo da cultura do milho; • Uso de folhas de eucalipto misturadas com gindungo no combate do gorgulho nos cereais e no feijão.	• Abertura artesanal de cacimbas em leito de cursos de água para assegurar o abeberamento animal; • Transporte manual de água até junto dos rebanhos; • Transumância sazonal em busca de pasto e água; • Deslocalização de rebanhos por via o okupakesa; • Alimentação do gado com restolho de culturas agrícolas.

Source: Ribeiro et al. (2021)

The region is very poor in modern practices and techniques in use in family agricultural and livestock production in the context of climate change mitigation. The table shows the main modern practices and techniques in use in family agricultural and livestock production in the context of climate change mitigation.

Table 2. Modern practices and techniques in use for climate change mitigation

PROVÍNCIA	PRÁTICA OU TÉCNICA MODERNA EM USO NA MITIGAÇÃO DE ALTERAÇÕES CLIMÁTICAS	
	AGRICULTURA	PECUÁRIA
CUNENE	• Participação do aproveitamento hidroagrícola de barragens edificadas pelo Estado (Calueque); • Construção de represas e implantação de valas de rega; • Adopção de variedades de milho de ciclo curto.	• Recurso a motos de três rodas e camiões-cisternas para dar de beber aos rebanhos;

Source: Ribeiro et al. (2021)

The Ministry of Territorial Administration and State Reform (2023), is the governing institution that establishes in its Article 30, line 1, that the Provincial Office of Agriculture, Livestock and Fisheries in Cunene is the service of the provincial executive, tasked with providing technical advice. to the governor in matters relating to agriculture, forestry, livestock and fisheries. Line 2 states that the same office has the following competences:

a) Promoting development policies for the agricultural, livestock and fisheries sectors, in conjunction with the province's local bodies;

b) To liaise with the province's local government bodies to implement policies aimed at promoting and developing the fisheries sector, its derivatives and seafood products, as well as ensuring the commercialisation and supply of salt and fisheries products to the province.

c) Promote the creation of efficient veterinary services, as well as phytosanitary and zoonoses surveillance and animal vaccination mechanisms at municipal and city level.

d) Promote the creation and conservation of parks, botanical gardens and zoos, ecological corridors and houses, forests, recreational patios, canals, catteries and the creation of municipal nurseries.

e) Promoting policies aimed at developing agricultural and livestock activities and the commercialisation of products derived from them.

f) Exercising other competences established by law or determined by higher authority.

In the author's opinion, with the conscious participation of the competent institutions and the fulfilment of their responsibilities, significant leaps and changes in production levels could be made in family, cooperative and business production plans.

III. MATERIALS AND METHODS

3.1. Location

The work was carried out between March and September 2023 in the Oshomucuyu locality, which is located in the Ondjiva commune, bordered to the north by a primary school, to the south by the Caxila III neighbourhood and to the east and west by zanas de chanas with two chimpacas that only have water during the rainy season.

3.2. Characterisation of the study area

The locality is characterised by generally poor soils, i.e. derived from rocks and lacking in nutritional elements; this was chosen from the reports made by the Ministry of Territorial Administration and State Reform (Ministério da Administração do Território e Reforma do Estado, 2010), which also states that the climate has semi-arid characteristics, marked by highly irregular rainfall and the frequent presence of drought events, which determines the intrinsic fragility of natural systems.

The regional moderado is structured by particular geomorphological units which are the chanas, the small rainwater basins, the relief forms that define the great natural and cultural value. The chanas are the main drainage system for the surface water that flows southwards as a result of the regional Cuvelai floods. The flood is the main guarantor of the surface water supply system, which gives rise to shallow groundwater.

3.3. Population and sample

The study population consisted of 15 peasant families from the town of Oshomucuyu, as all of them met the selection criteria of owning land for food production, ploughing crops and raising animals.

3.4. Methodology used

The research carried out is of the exploratory-descriptive type, since under these conditions, a diagnosis was made to find out about the conditions of the agricultural and livestock exploitation and technological innovation, in relation to the real conditions of the Oshomucuyu mines.

3.4.1. Methods and techniques

The research used theoretical methods such as:

Analysis - synthesis and historical - logic for the characterisation of the historical background of agricultural development in Cunene, in the theoretical references of local development, the impact of science and technology. Empirical techniques such as observation, interviews and research were also used to assess the current state of development of science and technological innovation in the town of Oshomucuyu. The non-experimental design was used as the empirical research method, with the use of scientific observation and research to capture information at a single point in time, with the aim of describing variables and analysing their incidence and interrelationship at a given moment, taking into account the criteria provided by Hernández and Baptista (2003). The modelling method was used to design action plans to help improve the development of science and technological innovation in the mines of Oshomucuyu.

3.4.2. Design methodological

The design was structured in phases that chronologically and systematically gave rise to the study's specific objectives. The research requires a holistic approach to its analysis, which integrates productive aspects with the efficient use of innovative technologies adapted to the conditions of the study area and their future implementation.

Phase I. Diagnosis of the state and utilisation of technological innovation in vegetable and livestock gardens in the community of Oshomucuyu

Visits were made to each of the properties, including informal and fairly flexible interviews and surveys to characterise the production systems in the community under study. The variables assessed were:

1. Technological innovation capacity
2. Agricultural activities carried out;
3. Use of techniques and technologies in carrying out agricultural activities;
4. Livestock farming and technologies

Identification of the limitations facing production systems using the SWOT Matrix Using the SWOT Matrix as a diagnostic tool, strengths and weaknesses were assessed, which together helped to diagnose the internal situation and gave us an overview of the reality of the territory studied.

Phase II. Elaboration of technological innovation actions in the conditions of the region, which influence the improvement of agricultural development in the locality of Oshomucuyu. Ondjiva.

This proposed action will benefit the rural community, the region's economy and the environment by combining traditional peasant knowledge with agricultural technology It is leaving behind the traditional use of the land and monoculture as a form of exploitation. All this under the principles set out by Giordano (2014) and Macieira (2017).

3.5. Analyses and processing of information

The research results were analysed using tables and graphs to make them easier to understand. The information was analysed using the Excel and Word packages in the Windows 10 operating system.

IV. RESULTS AND DISCUSSION

4.1. Characterisation of agri-production according to innovation capacity technological

Taking into account the results of the different diagnostic techniques used in the research process, it can be said that in the study area, family farming is the most predominant form of agriculture, and by the way, it is subsistence farming, where yields have been unsatisfactory to meet the families' expectations, the use of implements to carry out activities on the ploughs are rustic and maintain those learnt by family generations, always manually or with the use of animal traction in the smallest of cases; This shows that the producers have insufficient knowledge to apply more productive techniques and little capacity for innovation, based on the existing conditions in the production areas. The most developed agricultural activities on the outskirts of Oshomucuyo are related to vegetables, grasses, legumes and cucurbits; with the predominant crops being grasses (millet, massambala and maize); vegetables (cabbage, tomatoes, peppers, onions, aubergines, lettuce and carrots); cucurbits (cucumbers, pumpkins, watermelons and matanga) and legumes (fava beans, macunde beans). Livestock farming is highlighted as the municipality's greatest potential, with 450,000 head of cattle, which use free grazing as their only source of food. Most of the local producers use chemical fertilisers, while a smaller use organic fertilisers. The Agriculture Office, for the development of agriculture in the municipality, says it has provided training for farmers and sometimes offers seeds (millet, millet, rice, wheat and maize), all as part of the programme to reduce hunger and poverty.

4.2. Agricultural crops grown in the area study

The graph shows the agricultural crops most cultivated by the producers. 100 % of the sample reported growing vegetables, followed by 40 % and 20 % producing grains and fodder respectively, as well as 33 % growing other crops. There are no fruit or tuber crops, which could be very important for improving the family's bottom line.

It is noteworthy that horticultural species are found in 100 per cent of the ploughlands, which constitute the basic source of food for families in the region, along with massango and massambala, and in a smaller percentage grass as the only alternative for food. feed for livestock. According to Ottmann et al. (2010), when a producer sows several crops in his garden, this practice brings with it social benefits: the greater the demand in productive and/or quantitative terms, the greater the increase in production, the employment of labour and the consequent link to the market.

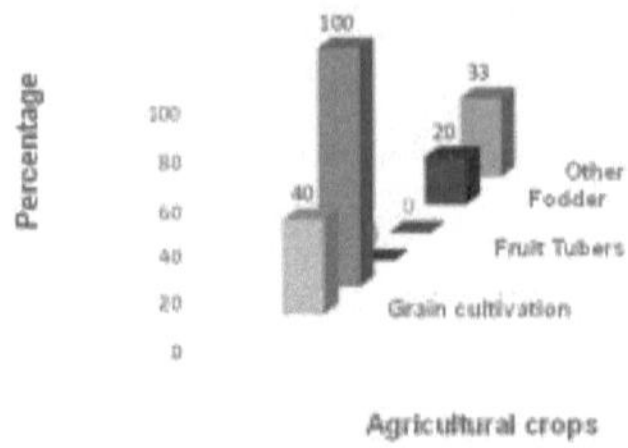

Figure 1. Agricultural crops most cultivated by producers

The Massango and Massambala do Sul Zone is characterised as an arid area of the country, with frequent droughts and prolonged dry spells. These conditions lead to very variable grain production and, in general, lead to a lack and/or reduction of pasture and surface water, which can result in significant losses of livestock, which is a key asset in the area (Capoulas, 2017).

Even though 0 % of the sample said they did not produce these crops, the survey found large fruit trees consumed by the inhabitants of the area. The lemon tree is the fruit for which there was the greatest diversification among the types planted, followed by papaya, guava and mango, although passion fruit is another of the most common. It should be emphasised that there were many trees bearing fruit typical of the region, but in none of the cases were these fruits used for processing and preservation for periods of scarcity; they were only used for consumption and the sale of surpluses.

As Capoulas (2017) said, in the context of the development of the population, especially the rural population, the promotion of family farming emerges as the basis of Angolan agriculture, fundamentally developed by peasants, who still practise farming systems using family labour, selling some of their surplus production to the consumer market, namely cassava, maize, beans, reindeer potatoes and also some vegetables and even fruit.

4.3. Use of techniques and technologies in carrying out agricultural activities

With regard to the technologies used by producers in the study area, the table shows that 100 % of them use some, with 73 % doing so manually, only 27 % using animal traction and 13 % using hoses for irrigation. It is very clear that there are shortcomings for producers in obtaining mechanised means to carry out agricultural activities, such as soil preparation and irrigation using electric motors or fuel, which makes the development of agriculture more precarious.

With regard to fertilisation, 100% of the sample reported using chemical fertilisers and 40% organic fertilisers, which indicates that even though this is a source of soil and water contamination and is more expensive, they still prefer to use chemical fertilisers, not taking advantage of techniques to make organic fertilisers earthworms, animal manure and crop residues. This is a topic that could be suggested for training, given the characteristics of the region and the benefits of organic fertilisers. With regard to the techniques used in agricultural management, 67 % use natural cover and 40 % of the sample uses leftover harvests, 33% practices monoculture and 27% intercropping, followed by 20% that uses polycultures. This shows that a large proportion of the population under study is unaware of the benefits of using techniques that, when used properly, increase crop production and, in turn, increase profits and improve nutrition.These results are very important as they can be compared with those of other authors such as Hugo et al. (2019), who state that cultivation can be conventional, with soil preparation, or direct planting. Conventional cultivation is more commonly used in the initial establishment phase, in the recovery of degraded pasture or in new areas, with the aim of incorporating correctives and fertilisers and better systematising the soil surface. The direct planting system (DPS) becomes more efficient after fertility correction and the adoption of soil conservation practices. In view of the benefits that the SPD provides, both annual crops and forage crops can benefit from it and be established in this planting system.

These results are also comparable to those of Melo (2016), who states that soil protection measures are also known as "palliative measures" in ecological, organic, biodynamic or agro-ecological agriculture and are an alternative for improving agricultural soils, proposing an appropriate management of the natural resources involved in production processes.

Table 3. Farming techniques and technologies used by producers

Indicators	Classification criteria	(%)
Utilisation e types of technology used	Yes	100
	No	-
	Manual	73
	Mechanised	-
	Animal traction	27
	Watering with a hose	13
	Other - Use of electric motor or fuel	-
Fertilisation	Yes	100
	No	-
	Chemistry	60
	Organic	40
Techniques used in agricultural management	Crop rotation	20
	Intercropping	27
	Monoculture	33
	Polycultures	20
	Natural cover	67
	Covering with crop residues	40
	Other - greenhouse	-

On the handling of rubbish (crop remains) in the backyard: "The bean straw is cut up and spread on the ground in preparation. He then collects the "fine" part of this partially decomposed waste and takes it to the vegetable garden beds. The "coarse" part is spread over other areas that are being opened up. This waste management is done only in the backyard area due to the quantity available and the distance from other plantations." Lobato et al. (2017) states that natural products can be used to enhance the ecological management of agricultural agro-ecosystems. On the management of guandu residues in the bean field: "Tie up the guandu residues, let them decompose and spread them with a hoe and harrow after the beans have been harvested".

In this respect, Arzola (2007) states that adequate doses of fertiliser protect the environment. This is possible by promoting more vigorous and productive crops with profuse root systems. In the more "bumpy" areas, planting is done in pits. In this way, he says he has been "surrounding the water with bananas". He chose the banana tree because "the root spreads out and prevents breakage, it filters.
Given the divergence observations, these are themes that deserve to be explored in the community's production systems. An example of this is that after harvesting, the plants are cut down and left in the ground, without burning the residue. In the third year, the maize is planted in the original bean rows or in its ratoon. Another rotation used is planting beans and maize in an area that was used to grow sweet potatoes the previous year; maize in succession to guandu and guandu in succession to white beans are also practices used.
All this confirms what Freitas (2011) says, who states that the role of beans in restoring fertility to the land is observed by farmers who grow them, and that there is great interest in cover crops that can be used as green manures.
He also argues that consortia are an object of study that should be prioritised, given that only one producer, out of 25% of the sample, does not use this system (and does not rotate either). Several divergent reports were made, so experiments on the species to be intercropped, spacing and planting times should be carried out. Among these reports, there is a success of a bean and maize consortium, provided that three rows of beans are planted to one of maize; another that informs about the viability of a bean and maize consortium only if the beans are planted days before the maize. In this sense, Carvalho et al. (2016) recommend the use of intercropping to control pests and increase beneficial insect biodiversity, which influences plant pollination.
According to Capoulas (2017), family farmers have to make the most of their land so that they preferably have a variety of foods throughout the year; there are societies in which family farmers monoculture in order to exchange production for other foods. Thus, they produce several species at the same time, which they can change according to the season, if that makes sense in regions where climatic variables remain stable all year round. Many of them, not having the resources to buy improved seeds or agrochemicals from the industry, preferentially choose, save and sow the seeds of plants that they consider most suitable in terms of production, resistance to disease or local soil and climate conditions. By doing this, they are practising a more "environmentally friendly" form of agriculture, protecting biodiversity, even without these express purposes. According to Hugo et al. (2019), several annual crops are used with crop-livestock integration in the Cerrado region, such as oats, millet, sorghum,

maize, soya, cotton, beans, wheat and, more recently, sugar cane. But the most prominent are soya and maize, due to their area, production and economic importance. The production systems used depend on the production focus and objectives of the property. On properties where the primary objective is to produce meat or milk, and where there is no interest in producing grain or preserved fodder due to a lack of equipment, the most commonly used annual crops are oats, millet or sorghum for grazing. In more complex systems, the same crops can be used, as well as others, such as maize and soya, which increase soil fertility to higher levels and provide conditions for the production of hay, silage or grain for animal production or commercialisation.

4.4. Livestock farming and technologies

As for the livestock technologies most used by the sample, free grazing is the only way of doing it, as confirmed by 100% of producers, and only 27% report having feed troughs and producing hay to preserve animal feed. This result reflects the fact that very few traditional techniques are used, without going into the usefulness of controlled grazing, dividing up the grazing area, creating protein banks, among others.

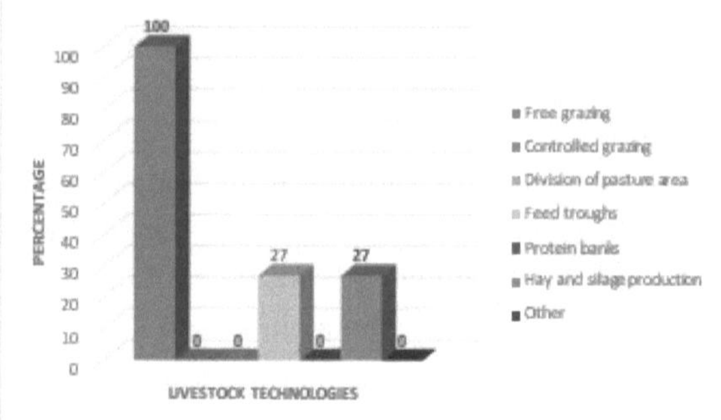

Figure 2: Livestock technologies used by producers

The properties visited have areas of surplus natural vegetation. As well as being located in areas that are difficult to exploit by farming activities, the farmers are aware of the importance of their preservation and regeneration and use them for the occasional collection of native fruit and let their cattle loose in these areas. These results coincide with those of Guimarães et al. (2000), who published that the family system has free-ranging animals that periodically move to favour fresh, new feed and sleep in the backyards or courtyards of the houses at night. According to Lopes et al. (2018), other widespread farming activities are sheep and goat farming, with herds of 66,895 head of goats and 110,294 head of sheep, most of which are raised extensively. Confined poultry shrank sharply, by around 50 %, while the other herds showed growth, particularly other poultry (450 %) and pigs, with around +90 %. The cattle, goat and sheep herds showed a lower variation than those, but a consistent one. As for the buffalo, donkey and mule herds, which are the least quantitatively expressive.

However, it can be said that the total headcount of these herds follows the trend found for the Northeast region (IBGE, 2008), where the sum of the equine, donkey and mule herds is not equivalent to 7% of the sum of the cattle, goat and sheep herds.

In the Brazilian semi-arid region, as in other developing regions, animal production stands out for its importance in food security, job and income generation, draught power, transport, and the production of fertiliser and fibre (Lima, 2006). Meat, eggs, milk and dairy products, together with plant products, can fulfil the nutritional requirements of human beings, as they are sources of lipids, proteins and carbohydrates (Torres et al., 2000). These animal production systems are related to production capacity, management, social benefits and even the survival of families in situations such as the periodic droughts that affect the north-east of Brazil. In this way, the diversification of production is generally related to financial conditions, the personal characteristics of the producer and the factors of animal production in the family farming economy, such as periods of water scarcity, soil aridity, distance from large input supply centres, among others (Bezerra et al., 2013).

4.5. Identification of the strengths and weaknesses facing production processes using the SWOT Matrix

Analysing the participatory diagnosis with the SWOT matrix (Table 4) revealed the existence of strengths, opportunities, weaknesses and threats that influence and determine the sustainable production process in the locality studied. We found a series of strengths and opportunities in the locality which, if harnessed, could reverse the existing situation through sustainable production on the farms, with benefits for the family economy, but specifically in improving agricultural management, availability of mechanisms and sources for the long-term acquisition of inputs, and improved prices as well as the availability of products in the nearest squares and informal markets.

Table 4: Strengths and opportunities in the production processes of the Oshomucuyu locality

STRENGTHS	OPPORTUNITIES
Producers' willingness to share experiences and learning. Presence of professionals in the locality, with the possibility of training. Availability of land (soils) and infrastructure.	The cultivable area can be increased. Income can increase due to the potential of available resources and innovative capacity. Financial support from the local administration to boost agricultural production. Existence of FRESAN non-governmental organisations spreading technical knowledge to promote sustainable agriculture and help fight hunger and poverty.

By determining the weaknesses and threats, we can offer alternatives for improvement, based on the principles of sustainable agriculture and technological innovation.

Table 4. Weaknesses and threats in the production processes of the Oshomucuyu locality**.**

WEAKNESSES	THREATS
Insufficient knowledge of production technologies and innovation alternatives for different crops. Poor use of soil improvement alternatives. Little use is made of crop by-products for organic fertiliser. Insufficient way of preserving food.	Decrease in public investment in the agricultural sector. Not everyone speaks and understands Portuguese. Frequent climate change. Not all farmers have the technological means and capacity for innovation.

To this end, more efficient ways must be devised to increase production by exploiting local resources and increasing the innovative activity of producers.

4.6. Proposal for technological innovation actions in the conditions of the region, which influence the improvement of agricultural development in the locality of Oshomucuyu. Ondjiva

The fundamental aim of these actions is to provide technical tools for the introduction and expansion of sustainable and resilient agriculture, for the improvement of organic and family farming and soil and water conservation in agricultural systems, based on the exploitation of the potential they have in their hands and their possibilities for innovation.

This proposal is based on the principles developed by researchers Giordano (2014) and Macieira (2017), adapted to the ethnic, cultural and economic realities of the area in question. It is organised in such a way that it responds to the most pressing needs of the population under study, but it is considered that other themes will be necessary according to the reality of the region.

Action 1.

Problem: insufficient knowledge of how to fertilise the soil.

There are six main elements that make up soil fertility. The big three are consumed in large quantities: nitrogen, phosphorus and potassium, and the other three are consumed in small quantities: calcium, sulphur and magnesium. We must add the trace elements, which are almost always found in sufficient quantities. The plant consumes them in very small quantities. Only analysis can accurately determine the deficiencies of a soil that is considered to be well supplied with between 1 and 2 per cent of the big three. However, whatever the richness of a soil, fertilisers are always needed as restitution.
Technological innovation: we propose three types of manure, depending on the quality of the soil.
Fertilising techniques to enrich the soil

• Corrective fertilisation: this is shock fertilisation, designed to bring the soil up to a suitable level. In addition to other analyses, here's how we assess this fertilisation.
In poor soils: In poor soils, weak contributions have practically no effect. They are energetically fixed by the soil. The plant can only benefit from them once it reaches a certain level.
It is the humic-clay complex that is formed under the influence of limestone, which regulates exchanges between the soil and plants.

1. Simple fertilisers: 100 g to 200 g of dried blood or pulverised horn, 60 g to 120 g of superphosphate, 50 g to 100 g of potassium sulphate.
2. In compound fertilisers: 200 g to 400 g of 3/6/9 fertiliser (or 150 g of 6/8/12)

• Investment fertilisation: this is an advance of fertiliser introduced into the soil. All fertilisers, except nitric, are well retained, so there is no risk of loss. This fertiliser is a safety net. It prevents the soil from falling below an excessively average level. We set it at 50 % of the corrective fertiliser.
• Conservation fertilisation: this is annual fertilisation to compensate for the

plant's surveys. These surveys are very different. But customised fertilisation would be too complicated, given the diversity of vegetables. Therefore, preference should be given to a maintenance fertiliser, which we could call a dressing, and partially corrected by the investment fertiliser: 100 g to 125 g of 3/6/9 fertiliser or 75 g to 100 g of 4/8/12 (or 6/8/12) fertiliser.

Other ways of supplying the soil with organic fertilisers are: Lumbrigaculture
It is a method that consists of breeding and cultivating roundworms in order to recycle the organic waste that is produced daily in farming. This industry alone requires minimal investment and infrastructure, and represents a major advance against environmental contamination and the problem of organic waste, as well as being an additional source of income through the concept of organic fertiliser (roundworm manure) and a source of protein for fish and poultry feed.
The infrastructure consists of interlocking thin round sticks filled with clay and coated with green cattle manure. The substrate is prepared with cattle, pig and sheep manure, which decomposes quickly, and some legumes native to the region, mixed in a ratio of 70 per cent legumes and 30 per cent cattle. It lasts approximately 45 days. The Californian red worm (Eisenia foetida species) can be used.
Composting

Composting, also known as the process of recycling organic waste, transforms the organic matter found in rubbish into natural fertiliser for use in agriculture. It is a natural process in which microorganisms are responsible for degrading certain types of waste, replacing chemical products and helping to reduce global warming.
The following steps must be followed to realise it:

1. Make rows separated by boards to isolate the compost space.

2. Dispose of crop residues and organic waste and place a layer of leaf litter or soil on top of this waste (there can be two or more layers).
3. Water every three days to ensure that the organic material decomposes.

4. Allow 45 to 60 days to harvest and guarantee the quality of the product.

We recommend creating several rows so that you always have plenty of compost to fertilise (one ends and another begins).

Action 2.

Problem: little use of soil conservation techniques.

Synopses: Soil conservation is necessary for several reasons: to combat harmful weeds, to aerate the soil and preserve its biology, and to limit water loss. It is practised with various tools: hoes, rollers, curved tines, hoes and ploughs. It is especially important when the surface of the soil takes on the consistency of a crust under the combined influence of rain and sun. The well-informed gardener should try to solve all these problems during the first year of cultivating their garden.

Technological innovation action: a preparation timetable is proposed to avoid mistakes in the different ways of tilling the soil.

Preparation calendar

1. From September to early January: Dig by hand or with a tiller, if the soil is well tilled.
2. From January to the end of March: weed by hand or with a power tiller, on plots that will receive crops at least six months later. Tilling is done without disturbing the soil if the crop is due to be cultivated. The work carried out must be followed by loosening and pressing the soil.
3. From the beginning of April to the month of August: only weed without turning.

Preparation techniques

✓ Labour by hand. Divide the plot in half. Make a starting pit, pouring the soil into the neighbouring half-plot, as this will be used to fill the last pit. Bury the hoe with one foot or your hands. Lift up the clod of earth and put it . But start by splitting the clod perpendicular to the pit, so that the you can remove them more easily. Coarsely crush the logs as you go.

✓ Preparation with a motorised tiller. With a simple plough, it's done in one direction only, with two working possibilities. Start in the centre and turn around.

✓ Divide the land into two parallel strips, go along one strip and come back along the other. With a two-mouldboard plough, make a first very straight furrow to adjust the depth. Make a second furrow, passing a wheel over the furrow to adjust the inclination of the plough's mouldboard. At the end of the furrow, lift the mouldboard, turn the tiller by unhooking the left wheel to go to the left (or the other way round), make a quarter of a turn going forwards and unhook the right wheel to make a quarter of a turn going backwards; hook the two wheels, turn the mouldboard and start a new furrow.

✓ Tillage without turning. With a hoe or a narrow spade: get into position to weed by hand, but without making furrows. The technique consists of burying the hoe, lowering the handle as low as possible, shaking it and starting again at the side. With a special hoeing tool, the technique is the same, but the shape of the tool (two handles or T-handles) allows you to shake the clod sideways rather than vertically.

✓ Tilling with a mechanical hoe. Starting from the edge of the separation furrow, lower the spur to attack the first strip, with the plough's tines wide apart, so as not to go over the part already worked. At the other end, raise the spur and turn round to the second strip, burying the spur and moving the tines to the other side. Finish with an uncut.

✓ Tillage with a double pit. It's interesting in vegetable gardens - when you want to increase the thickness of the arable layer - and on a prairie, when breeding. It is a double-dig job, thanks to the establishment of a furrow at two depths. When forming the second furrow, the top clod is drawn to the bottom and the bottom clod to the top. The second furrow is then formed. On large plots of land, practise deep ploughing with a plough if you want to respect the order of the layers.

Action 3.

Problem: little use of techniques to increase yields.

Synopsis: Striving for productivity is a reasonable and even necessary desire. But good techniques are needed, and these are all based on the data of intensive cultivation and rotations. Growing vegetables denotes a positive intention: the desire for fresh, healthy crops whose origin is trustworthy. This is possible with good cultivation techniques.

Technological innovation action: proposal of different cultivation techniques and protection against pests and diseases.

✓ The construction site and its position

The jardiniere is indispensable in all vegetable gardens. It is a terrace located next to a wall or a wall that is well exposed to the sun (south, south-east or south-west). It is sloped to maximise the sun's heat. If you don't have a , slope the surface of a board towards the ground; the soil on the opposite side will be held in place by boards or a second slope. In this case, the bed is called a slope. In a bed, you can grow first fruits, lettuce and other salad plants, and produce all the seedlings you need for transplanting.

✓ Crop rotation

This is the succession of crops in a given garden bed. It is not to be confused with crop rotation, which covers all the rotations in plots or "fallows". A well-organised vegetable garden is divided into four plots. One of them is reserved for perennial vegetables, which occupy this space for years (leafy cabbage, gindungo, ginger, aromatic plants, among others). In the other three plots, alternate the vegetables. As shown in the table.

Table 4: Examples of crop rotation by plot

Years	Plot 1	Plot 2	Plot 3		
1O year	Potatoes, carrots, turnips, tomatoes, beetroots	Cabbages, endive, spinach, lettuce	Beans, broad beans	peas	e
2nd year	Beans, peas, broad bean (legumes - seeds).	Potatoes, carrots... (root vegetables and fruit).	Kale, endives... (leafy vegetables).		

✓ Intensive cultivation

Intensive cultivation allows you to harvest the maximum amount of vegetables on a minimum area. It requires methodical cultivation, sensible use of manure and succession of crops with the use of hidden crops, counter planting, counter sowing and combined sowing. During the season, no part of the garden should be left uncultivated for more than 24 hours.

✓ Greenhouse

When using greenhouses, adapting to your environment and using techniques that simplify work increase yields. This process of artificially regulating heat is essential in all good gardens, especially small ones. As well as light, a greenhouse needs ventilation so that the place doesn't get too hot and fulfils its purpose.

They are fixed or removable. The support is made of wood and the cover is made of rigid polyester. They are installed in a well-exposed location. The gardener can obtain products from the seedlings to be replicated (cabbages, tomatoes, aubergines, peppers, watermelon...).

✓ Vertical cultivation

In small gardens, where the horticulturist struggles with a lack of space, the technique of growing on stakes or nets is almost a necessity. On a given, generally small area, it multiplies yields by 3 or 4. In the past, there was the problem of stakes. Today, it has disappeared thanks to climbing plant netting, made with meshes from 4 cm to 10 cm on a side. So you can grow beans, peas, melons, pickling cucumbers and salad cucumbers on cuttings too. Not to mention vegetables whose cuttings are almost obligatory, such as tomato plants.

✓ Traditional watering and tricks

Apart from the watering can and the hose used to water small seedbeds, the best way to water a vegetable garden is certainly with the oscillating watering can. place it on the edge of the knowledge and adjust it at will, while it's running and without risk of getting wet.

Some of the watering methods we call soaking make it possible to keep the soil moist locally, at root level. They are particularly interesting for melons, pickling cucumbers, salad cucumbers, pumpkins, celery, tomatoes, aubergines and peppers.

- The bottle tilted towards the plant: Wet the area well, make a mould of the bottle half buried at 45 degrees. Fill it with water and put it back in place.
- The jug between two plants: This is watering in case you have to be away for a long time. The 51 to 201 jug is buried up to the shoulders, filled to 4 /5. The expanding air pushes the water out during the day and recovers it at night.
- Underground watering by perforated tube or porous hose, which eliminates the need for dripping water.

✓ Conservation

It's done in various ways.

- Preservation in the ground, under cover. This is a practical process when the soil is not damp and applies to root vegetables. Their foliage is cut off at the beginning of December. They are covered with a mixture that can be lifted with a crepe and allows the vegetables to be harvested from soil that remains soft.
- Store in a dry place. Good for seed vegetables, fruit vegetables and tuberous bulbs.

Seed vegetables. Then fix and hang them up out of the rain.

Fruit vegetables. Unripe tomatoes, aubergines and peppers placed on mobile shelves will keep for 2 to 3 months. Squash and heads will keep for a long time if they haven't been opened; if they have, soak the cut side in salted water.

Bulb vegetables. Keep them hanging in small bundles (onions), in braids (garlic), or arranged on shelves.

- Preservation by drying. Dry peeled peas, carrots, turnips and onions in the oven, cut into slices and used to flavour soups. Beans can also be preserved by stringing them and drying them in a dry place.
- Preservation in pits or greenhouses. This process is used for cabbages and salads. They are placed next to each other, with their respective clods of earth, in a hole made in the ground and covered with a layer of straw. You can also store them in greenhouses.
- Storage in silos. This refers to root vegetables that have been harvested with their stems cut and well-dried. They are stacked in pits in the ground, or even on top of the ground, covered with straw, earth or even plastic to prevent rainwater from penetrating.
- Store in the cellar. Useful for a short period of time for preserving endives with their clumps. Also suitable for storing root vegetables in stratifications in the middle of dry sand. Store your potatoes in the corner, insulated from the ground by a board, in a box with a crate or in a metal silo.

✓ Sustainable pest management

The poor management of pests in crops is the result of successive applications of non-selective chemical products to control them, which is why they have acquired resistance, causing a reduction in yields, irreparable damage to human health and environmental contamination. This, coupled with rising production costs, highlights the need for new alternatives that are sustainable and low-cost.

- Plant preparations are recommended, which are plant extracts that contain certain active components which are effective if applied to the soil or crops as activators of soil life, to vigorise it, or to manage pest or disease problems, using them as insecticides, fungicides, fungistats that stop the growth of mushrooms and as repellents. Examples of this are natural insecticides such as tobacco (Nicotina tabacun), chilli peppers (Capsicum frutezco), and others that help reduce the cost of agrochemicals and, in turn, provide efficient management and help conserve the environment and the beneficial fauna within cultivated fields.

As an example of how to treat pests with tobacco extract, 2 kg of dried tobacco is chopped and boiled in 5 litres of water for 15 minutes, left to stand in the shade and then water is added to make up to 20 litres.

Action 4.

Problem: insufficient management of animal production systems

Synopses: livestock production systems are characterised by herds of numerous animals, extensive exploitation, natural pastures with low levels of technology, low availability of food during the dry season, poor pasture management due to overgrazing with a racial component of mixed-breed (native) animals and low levels of milk and meat production.

Technological innovation action: it is proposed that technological alternatives include management practices, which are understood as the rational use of the property's available resources in order to obtain greater production per unit area.

Bovine and/or sheep-goat production system

✓ Division of grazing areas: allows for better management of animals and rational use of grasslands to develop these practices (select an area with three divisions to carry out the rotational quartering system).

✓ Feeding alternatives: Adapt an alternate quarter, trying to preserve natural silvo-grazing conditions using the existing forage resources on the property, especially native legumes and sugar cane, to be used in critical times of drought.

✓ Animal improvement: Selection and introduction of a group of animals from improved breeds (cattle, goats), development of appropriate management, health, feeding and nutrition practices.

Pig production system

✓ Utilisation of by-products from crop residues.

✓ Sustainable alternatives that improve production with improved breeding stock, with an adequate health plan and feed (wheat, maize, cane (energy source), whey, native legumes (protein source)).

✓ Construction of a wooden pigsty with breeding rooms, a farrowing room, a zinc roof and cement floor, drinking and feeding troughs in each room, an area for remaining pigs and those of breeding age, as well as setting aside a neighbouring block for grazing with native grasses and legumes.

Poultry production system

✓ Construction of a half-water coop 2.5 metres high at the back and 2 metres at the front, with a tiled roof and wooden walls and drainage channels. With drinking fountains, feeding troughs and nests, a neighbouring corral with divisions and sown with pasture and legumes native to the area.

Biogas plant

As a mechanism for recycling and treating organic waste from animals, you can install a biogas plant (different forms), which needs to have the following parts.

1. Mixing tank where the manure is mixed with water in a 1:1 ratio, one bucket of water for one of green manure.
2. Biodigester, which is a digestion chamber where the raw material to be fermented is deposited.
3. Gasometer or gas tank, joined to the biodigester or forming a separate unit
4. Compensation or leakage tank, where the effluent or bio- fertiliser is received.

This product (gas) is used to cook food and the effluent or bio-fertiliser is used to fertilise vegetables and fruit trees.

4.6.1. Expected results

By using this proposal for actions, families can guarantee their own food supply and generate a surplus that can be sold to meet their needs. In addition to having the advantages that make their gardens more productive, such as:

1. Rational use of family labour throughout the year.
2. Better utilisation of the soil and reduction of the effects of crop season.
3. Increased production per unit area.
4. The use of chemical fertilisers is reduced and organic farming is promoted.
5. Increasing animal production and its benefits.

V. CONCLUSIONS

1. Through the application of research methods and techniques, it can be confirmed that producers in Oshomucuyo practise traditional agriculture and livestock breeding and that the technologies used do not reach the level of innovation.
2. Strengths, opportunities, weaknesses and threats were identified using the SWOT Matrix.
3. A proposal of actions has been drawn up, the fundamental aim of which is to provide technical tools for the introduction and expansion of sustainable and resilient agriculture in the locality of Oshomucuyo. Ondjiva.

VI. RECOMMENDATIONS

1. Putting the proposed system of actions into practice to confirm its objectivity in the locality of Oshomucuyu. Ondjiva
2. To disseminate the results of this work in educational centres in the town of Oshomucuyu and other towns in Cunene province, with the participation of the relevant institutions.

VII. REFERENCES BIBLIOGRAPHIC

Arocena, J. (1997). "Local development in the face of globalisation", Universidad de Chile

Astier, M., Masera, O., and Galván, Y. (2008). Sustainability Assessment. A dynamic and multidimensional approach. Available at : http://docplayer.es/14885206- Evaluacion-de-sustentabilidad-un-enfoquedinamico-y-multidimensional.html.

Bohórquez, H. (2003). Ongoing Training, Something More than Simple .

Available at www.gestiopolis.com

Capoulas, L. S. (2017). Family Farming and Sustainable Development in the CPLP. Minister of Agriculture, Forestry and Rural Development of Portuguese Government. Minister of Agriculture, Forestry and Rural Development of Portuguese Government.

Carvalho, A. H. de O., Delprete, S. I., Costa, A. C., Lima, W.L. and Oliveira, F. L. (2016). Agronomic and economic viability of lettuce and spring onion intercropping. Cadernos de Agroecologia.

Carvalho, H. G., Roberto dos Reis, D., and Cavalcante, M. B. (2011). Innovation management.Curitiba: Aymará, (UTFinova Series).

Castro, F. (2002). "Science, Innovation and the Future". Cuban Book Institute. La Habana. Cuba. p. 189

CEIC/UCAN.(2019).REPORT ECONOMIC ANNUAL OF ANGOLA 2018,Catholic University of Angola, Luanda.

CEIC/UCAN. (2021).REPORT ECONOMIC ANNUAL OF ANGOLA 2019 Catholic University of Angola, Luanda.

CEIC/UCAN. (2021) REPORT ECONOMIC ANNUAL OF ANGOLA 2019, Catholic University of Angola, Luanda.

Escobar, C. (2010). Nature and impact of innovation in manufacturing industry: evidences from the Italian innovation survey. Res. Policy. 26:521

Franco, R. (2000). La agenda social de América Latina al comienzo del tercer milenio y el papel de las políticas sociales, Santiago, ECLAC.

Freitas, K. P. (2011). Production and appropriation of urban space in Campos dos Goytacazes

- RJ: from single-family homes to flat blocks. Master's thesis-UENF Gliessman, S. R. (2000). Field and laboratory investigations in agroecology. Boca Raton: CRC Press LLC,. 330 p.

Giordano, L. (2014). THE GARDEN - Varieties, Culture, Production. Publicações EUROPA- AMÉRICA, LDA APORTADO 8 2726 MEM MARTINS CODEX PORTUGAL.Issue [No] 136508/6065
Guimarães Filho, C., Soares, J.G.G. and Araújo. G.G.L. (2000). Goat and sheep meat production systems in the semi-arid Northeast. In: International Symposium on Meat Goats and Sheep, 1, 2000, João Pessoa. Proceedings. João Pessoa. p. 21-33.

Hernández, S.R., Colinas, C. and Baptista, P. (2003). Research Methodology, 684pp.

Hirakuri, M. H. (2012). Production systems: concepts and definitions in the agricultural context. Documentos, Londrina, n. 335, 2012. http://scielo.sld.cu/scielo.php?pid=S086439420070003000008&script=sci_arttext &tlng=pt
Hugo, Z. A., Motta, M. M. Cl., Volpe, E., Neivo, K. A. and Madureira, B. I. (2019). Crop management in crop-livestock integration

INE, (2018). Inquérito Integrado sobre o Bem-estar da População - 2010, Ministry of Planning, Luanda.

Lal, R. (2004). Carbon emission from farm operations. Environment International 30: 981- 990. IPCC. 2001. Climate change: the scientific basis. Intergovernmental Panel on Climate Change. Cambridge Univ. Press. Cambridge, UK.

Lima, G.F. da C. (2006). Alternatives for the production and conservation of strategic forage resources in the northeastern semi-arid region. In: ENCONTRO NACIONAL DE PRODUÇÃO DE CAPRINOS E OVINOS, 1, 2006, Campina Grande. [Proceedings]. Campina Grande: SEDAP, 2006.

Lobato, G. de J. M., Lucas, F. C. A., Tavares-Martins A. C. C., Jardim M. A. G. and Martorano

L. G. (2017). Diversity of use and socio-environmental aspects of urban backyards in Abaetetuba, Pará, Brazil. Brazilian Journal of Agroecology, 12(2). ISSN 1980-9735

Lopes, S. Y., Gamarra-Rojas, G., Paiva, F. F., Farias, J. L. de S. and Fernandes, C. de S. (2018). Livestock production in the family farming economy: a case study in the Brazilian semi-arid region. Cadernos de Ciência & Tecnologia, Brasília, v. 35, n. 1, p. 53-74, jan./abr. 2018.

Lote, E. R. (2015). Agricultural enterprises and economic development: Potentialities of huambo province. Portucalence University.

Macieira, M. M. V. (2017). "Feasibility project for the implementation of an integral property in the Francisco do Orellana canton, for the year 2015" Thesis prior to obtaining the title of Commercial Engineer. SHOP - Ecuador.

Marassiro, M. J., Romarco de Oliveira, M. L., and Pereira, G.da P. (2021). Family farming in Mozambique: Research, Society and Development, v. 10, n. 6, e22110615682, 2021 (CC BY 4.0) | ISSN 2525-3409 | Available at: http://dx.doi.org/10.33448/rsd- v10i6.15682

MARE, Ministry of Territorial Administration and State Reform (2010). General characterisation of Cunene Province. Available at http/infoangola.ao/index. Php.

MATRE, Ministry of Territorial Administration and State Reform (2010). General characterisation of Cunene Province. Available at http/infoangola.ao/index. Php.

Melo, L. P. (2016). The benefits of urban and peri-urban agriculture for the sustainability of the city of Macapá-AP. Pluris- CONGRESSO LUSOBRASILEIRO para o Planeamento Urbano, Regional, Integrado e Sustentável.

MINADER. (2020). 2019/20 Agricultural Campaign Report. MINADER. (2022). Report on the 2020/21 Agricultural Campaign. Ministry of Territorial Administration and State Reform (2023).

Neto, J. (2008). Angola: Agriculture and Food. Portuguese Institute for Development Support, Documentation Centre. Lisbon, Portugal.

Nevado, D. (2003). Management Control Renewed. Factor Humano y Nuevos Instrumentos de Gestión Empresarial. AECA Monographs.

Núñez, J. (2019). IV International Conference "The work of Charles Marx and the challenges of the 21st century" UNAH

Ottmann, M. M. A., Borcioni, E., Mielke, E. and Da Cruz, M. J. R. (2010). Environmental and socio-economic impacts on community gardens under transmission lines in the Tatuquara neighbourhood, Curitiba, PR, Brazil. Revista Brasileira de Agroecologia, v.5 (1), p. 86-94.

Pedro, H. A. (2008). Public Irrigation in Angola: Proposal for a Management

Model. Final Course Work in Agronomic Engineering. Higher Institute of Agronomy. Technical University of Lisbon, Lisbon. 50 pp.

Pérez, R. (2003). Proposal for a Human Management Model for Innovative SMEs.

Revista Escuela de Administración de Negocios No. 47.

Pineda, A. (2006). Valores. Cuban Managerial Issues. CCED, MES

Prevedel, D. (2023). Agricultural production: understand the types and systems of production. Available at: https://blog.aegro.com.br/producao-agricola/.

Republic of Angola (2020). Fundamental Report of the General State Budget (OGE 2020).

Ribeiro, C.M.F., Satiaca, C.S. and António, F.P.L. (2021). Practices and techniques in use in family farming and pastoralism in the provinces of Cunene, Huíla and Namibe Instituto Superior Politécnico Tundavala, Tchitoco area, Huíla, Angola: Department of Engineering.

Rocha, C., Mora, J. and Romero, J. (2016). Typology of production systems in the rural area of the municipality of Ibagué, Colombia. Agronomía Mesoamericana 27(2), 253-264. doi: 10.15517/am.v27i2.2436

Simões, J. (2019). Commercialisation as a factor of social promotion in Angola, Edition of the Commercialisation Fund, Luanda.

Socorro, O. (2003). ¡PROHIBIDO INNOVAR! I: Cuando el Cambio No Es Más que Una Palabra. Available at: www.gestiopolis.com

Suárez, J. (2007). Factores de éxito y fracaso en el comportamiento innovador de las empresas ganaderas cubanas. Available at

Torres, E.A.F.S., Campos, N.C., Duarte, M., Garbelotti, M.L., Philippi, S.T. and Minazzi- Rodrigues, R.S. (2000). Centesimal composition and calorific value of foods of animal origin. Food Science and Technology, v.20, p.145-150, 2000. DOI: 10.1590/ S0101-20612000000200003.

Tushman, M.L. & O'Reilly III, Ch. A. (2002). "Winning through innovation. A practical guide to leading organisational change and renewal". Harvard Business School Press, Boston.

Victorino Mauro, A. C. B. (2018). Case study in the agricultural sector in Angola: applicable accounting practice and proposal for its improvement. ASSOCIATION OF NORTHERN POLYTECHNICS (APNOR). BRAGANÇA POLYTECHNIC INSTITUTE

VIII. ANNEXES

Annex 1. Peasant survey guide.

Dear producer, in order to characterise the production systems in place at your mine and the implementation of technologies in your processes, the following survey is being carried out. We hope that the criteria you provide will enable us to make proposals to improve the situation in the area.

Sex , Age, Educational level of the head of.

Questionnaire

1. What agricultural activities do you carry out the most?

Growing grains, vegetables, fruit, tubers, fodde , livestock others

2. Do you use any technology in your farming activities? Yes, no .
How it's done: Manual, mechanised, with animal traction other

3. Do you use fertiliser in agricultural production? Yes

Which ones?

4. Crop rotation technique, intercropping, monocultur, polycultures, natural cover, cover with crop residues, others
5. Do you keep livestock? Yes

6. Do you use any technology in your livestock activities? Yes, no .
How much?

Free grazing, controlled grazing, division of the grazing area feed troughs banks , hay and production other

THE END...

Thank you very much for attention.

Appendix 2. Interviews conducted with the Municipal Director of Agriculture and Forestry, Mr Alfeu Haidudu Haikea, and the advisor to the office, Mr Ernesto Daniel Langesapwaka.

Dear friends, with the aim of characterising the production systems in the town of Oshomucuyo and the implementation of technologies in their processes, we are conducting the following interview. We hope that the criteria you provide will enable us to make proposals to improve the situation in the area.

1- What type of agriculture predominates in the municipality, especially in the town of Oshomucuyo?
2- What has your income been?

3- Have farmers used or implemented other techniques?

a) What is lacking in other farmers, is it a lack of mastery, knowledge or their own means?
b) What are the most activities?

4- Looking at agricultural fields and farmers, have they practised crop rotation?
5- What are the main crops grown in the ?

6- What is the municipality's great potential?

7- How has shepherding been?

8- How has the soil been prepared?

9- Fertilisation

10- What has the Agriculture Office done to develop agriculture in the municipality?
11- How many farmers are there in Ondjiva?

Thank you very much

Annex 3. Production areas with evidence of poor agricultural management and insufficient use of technological innovation

Annex 4. Free grazing system in unproductive and undelimited areas

MIX
Papier aus verantwortungsvollen Quellen
Paper from responsible sources
FSC® C105338

Printed by Books on Demand GmbH, Norderstedt / Germany